Arab Public Library
20020304

D1523365

PAPER HORNETS

ROSS E. HUTCHINS

PAPER HORNETS

illustrated by
Peter Zallinger

Addison-Wesley

Addisonian Press titles
by Ross Hutchins
THE CARPENTER BEE
THE CICADA
THE MAYFLY

An Addisonian Press Book

Addison-Wesley Publishing Company, Inc.
Reading, Massachusetts 01867
Printed in the United States of America
Second Printing

HA/BP 02986 8/74

Library of Congress Cataloging in Publication Data

Hutchins, Ross E
Paper hornets
SUMMARY: Describes the characteristics and habits of the hornet that utilizes special paper-making cells in constructing its nest.
"An Addisonian Press book."
1. Paper wasps—Juvenile literature. (1. Paper wasps. 2. Wasps)
I. Zallinger, Peter, illus. II. Title.
QL568.V5H84 595.7'98 72-4789
ISBN 0-201-02986-3

In the Great Smoky Mountains there is a quiet valley. Surrounding it are high, forested mountains. Numerous small streams of crystal-clear water flow down the steep slopes of the mountains, dropping over ledges and into quiet pools. All these streams unite, at last, forming Little River which, in turn, rushes down the valley and out of the mountains through a beautiful canyon-like gorge.

Our story begins in spring, after a long, hard winter. Here and there in the forest, beneath the bark of some fallen logs, are hornet queens who found refuge there in autumn. During the long, cold months they have hibernated in tiny cells, protected from enemies and weather.
Now, warmed by the sun, the queen hornets crawl out of their cells and fly away through the forest. Winter is past and they must build nests and start new hornet colonies in the forest.

First, however, the queens must find food after their long fast. They visit early flowers blooming on the mountainsides and drink their nectar. After feeding for a few days, they fly about through the forest seeking suitable places to build nests. Some of them fly up the valley, others go down the river. A few of the queens are captured and devoured by hungry birds. There are many enemies ready to to destroy them.

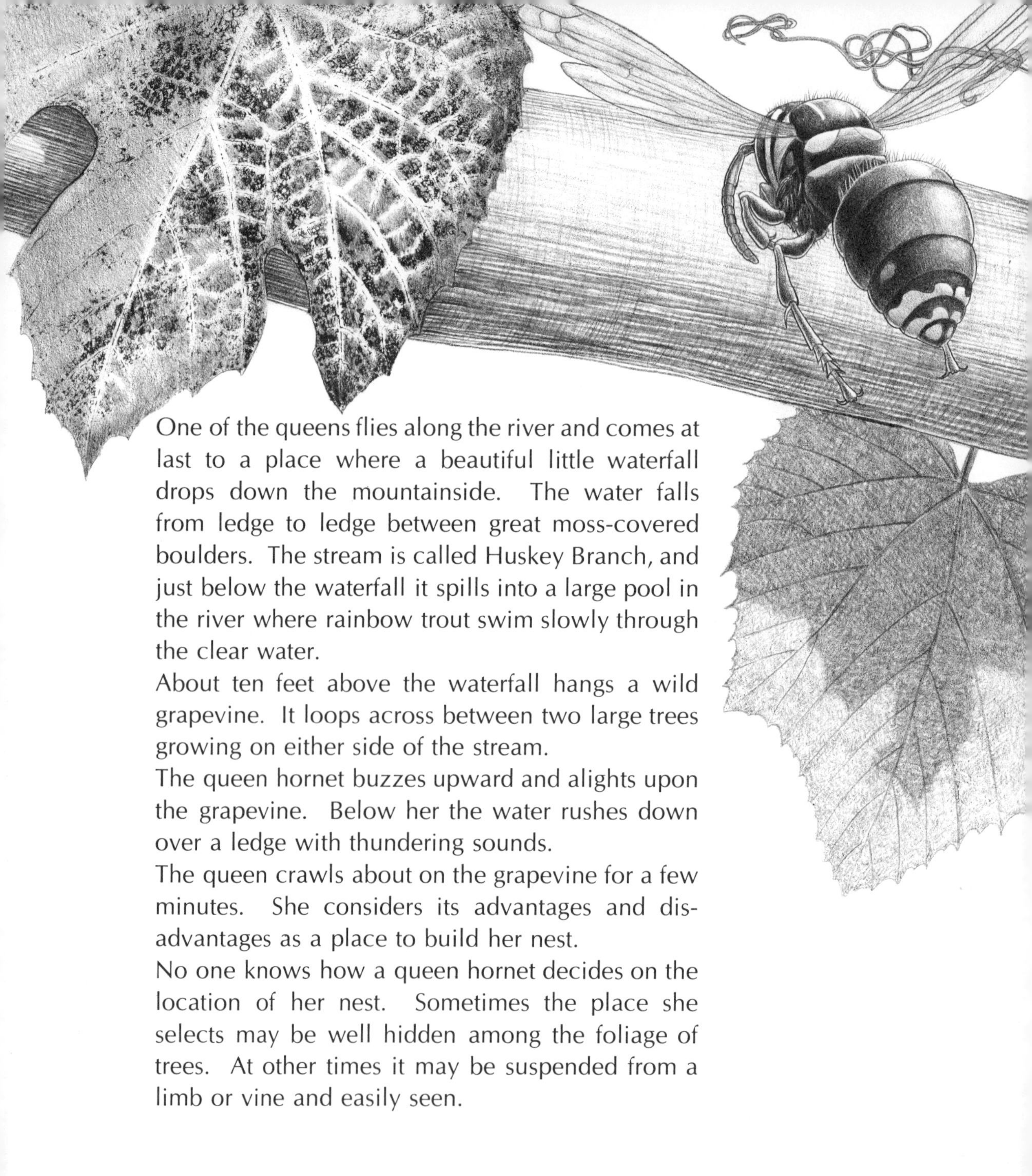

One of the queens flies along the river and comes at last to a place where a beautiful little waterfall drops down the mountainside. The water falls from ledge to ledge between great moss-covered boulders. The stream is called Huskey Branch, and just below the waterfall it spills into a large pool in the river where rainbow trout swim slowly through the clear water.

About ten feet above the waterfall hangs a wild grapevine. It loops across between two large trees growing on either side of the stream.

The queen hornet buzzes upward and alights upon the grapevine. Below her the water rushes down over a ledge with thundering sounds.

The queen crawls about on the grapevine for a few minutes. She considers its advantages and disadvantages as a place to build her nest.

No one knows how a queen hornet decides on the location of her nest. Sometimes the place she selects may be well hidden among the foliage of trees. At other times it may be suspended from a limb or vine and easily seen.

In any case, our queen at last decides to build her nest on the grapevine above the falls of Huskey Branch just before it flows into Little River. The beauty of the surroundings, of course, does not enter into her considerations. Probably she does not even notice the lovely waterfall or the way the droplets of water splash upon the nearby violets and ferns.

Having made her choice of a place to build her nest, the queen then flies to a dead pine a short distance away and comes to rest upon a limb from which the bark has dropped away. Here she scrapes up a mouthful of the gray fiber and chews it into a soft pulp.

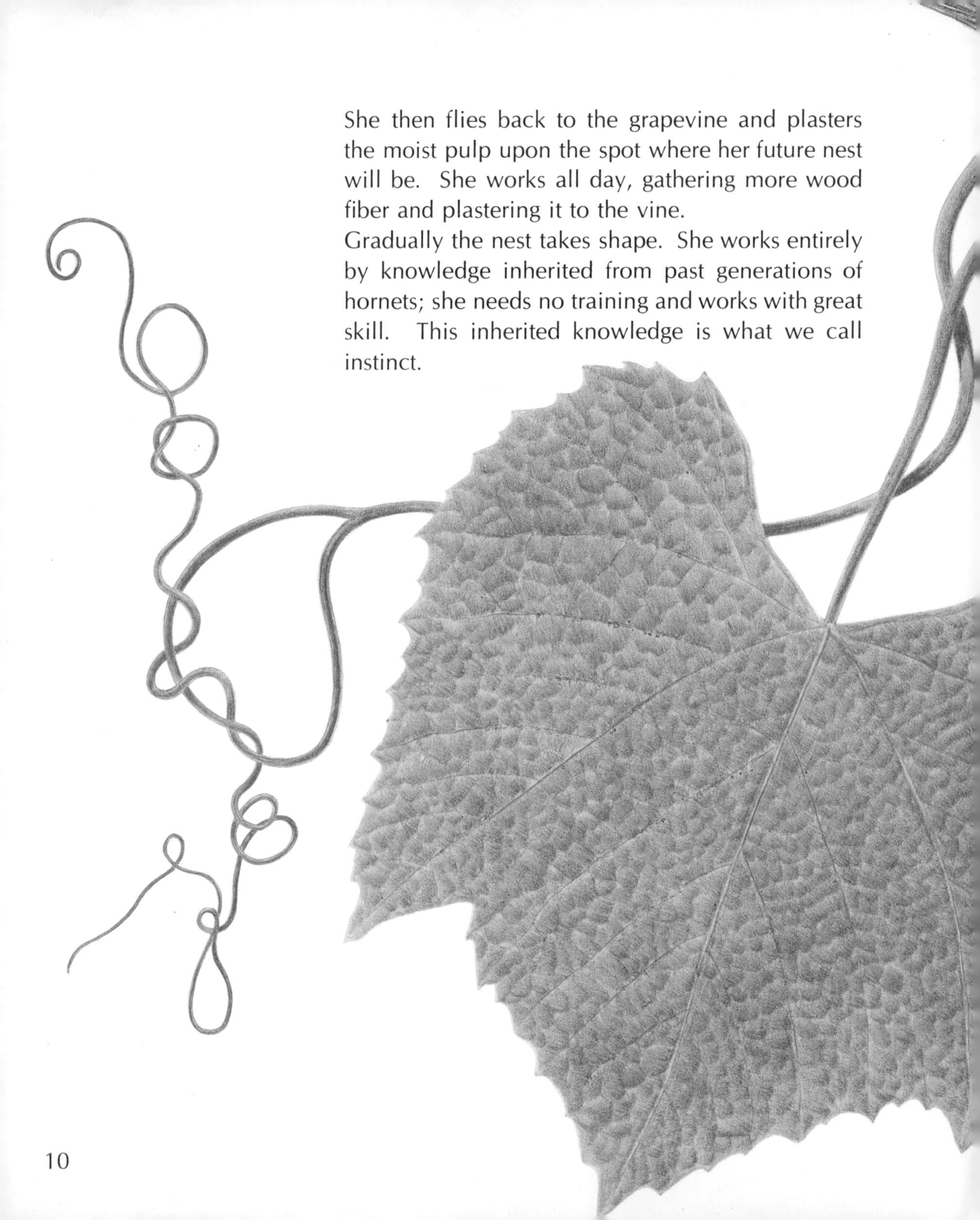

She then flies back to the grapevine and plasters the moist pulp upon the spot where her future nest will be. She works all day, gathering more wood fiber and plastering it to the vine.

Gradually the nest takes shape. She works entirely by knowledge inherited from past generations of hornets; she needs no training and works with great skill. This inherited knowledge is what we call instinct.

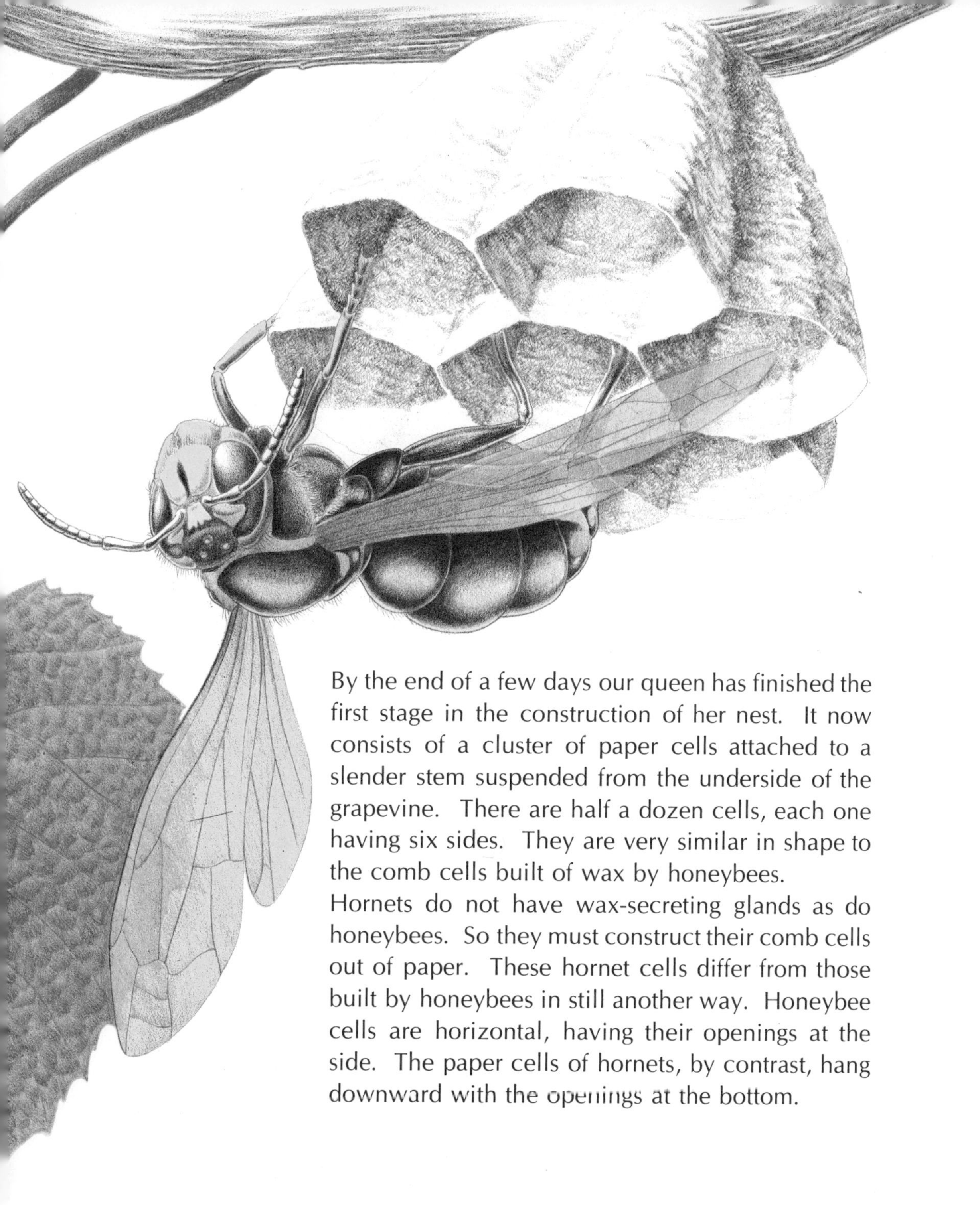

By the end of a few days our queen has finished the first stage in the construction of her nest. It now consists of a cluster of paper cells attached to a slender stem suspended from the underside of the grapevine. There are half a dozen cells, each one having six sides. They are very similar in shape to the comb cells built of wax by honeybees.

Hornets do not have wax-secreting glands as do honeybees. So they must construct their comb cells out of paper. These hornet cells differ from those built by honeybees in still another way. Honeybee cells are horizontal, having their openings at the side. The paper cells of hornets, by contrast, hang downward with the openings at the bottom.

Having finished her cluster of cells, the queen wasp gathers more wood fiber and begins building a sheath or envelope to enclose them. She chews the fiber in her jaws forming it into thin sheets that resemble paper.

Gradually the covering of the nest takes form. After a week of hard work the queen's nest is finished. It is globe-like, about the size of a golf ball, and open at the bottom.

She next lays a white egg in each cell, cementing it to the bottom with glue from special glands. For the moment, her work is finished. Her queen nest has been built and the first eggs have been laid in the cells.

She must now wait for the eggs to hatch.

For several days the queen mother remains within her nest, guarding it from enemies. Now and then she flies down to drink nectar from spring flowers blooming along the stream. A few times she alights upon the moss beside the waterfall and drinks water from droplets splashed up by the falling water.

Several times she has narrow escapes from birds. Once a jay darts down and almost catches her but she flies away just in time. Another time, as she sips nectar from a flower, a yellow crab spider grabs her, but she escapes.

Everywhere in the forest there are enemies that live by capturing and eating hornets and other insects.

After about a week the eggs laid in the paper cells hatch into tiny white larvae or grubs. The queen must now begin feeding them.

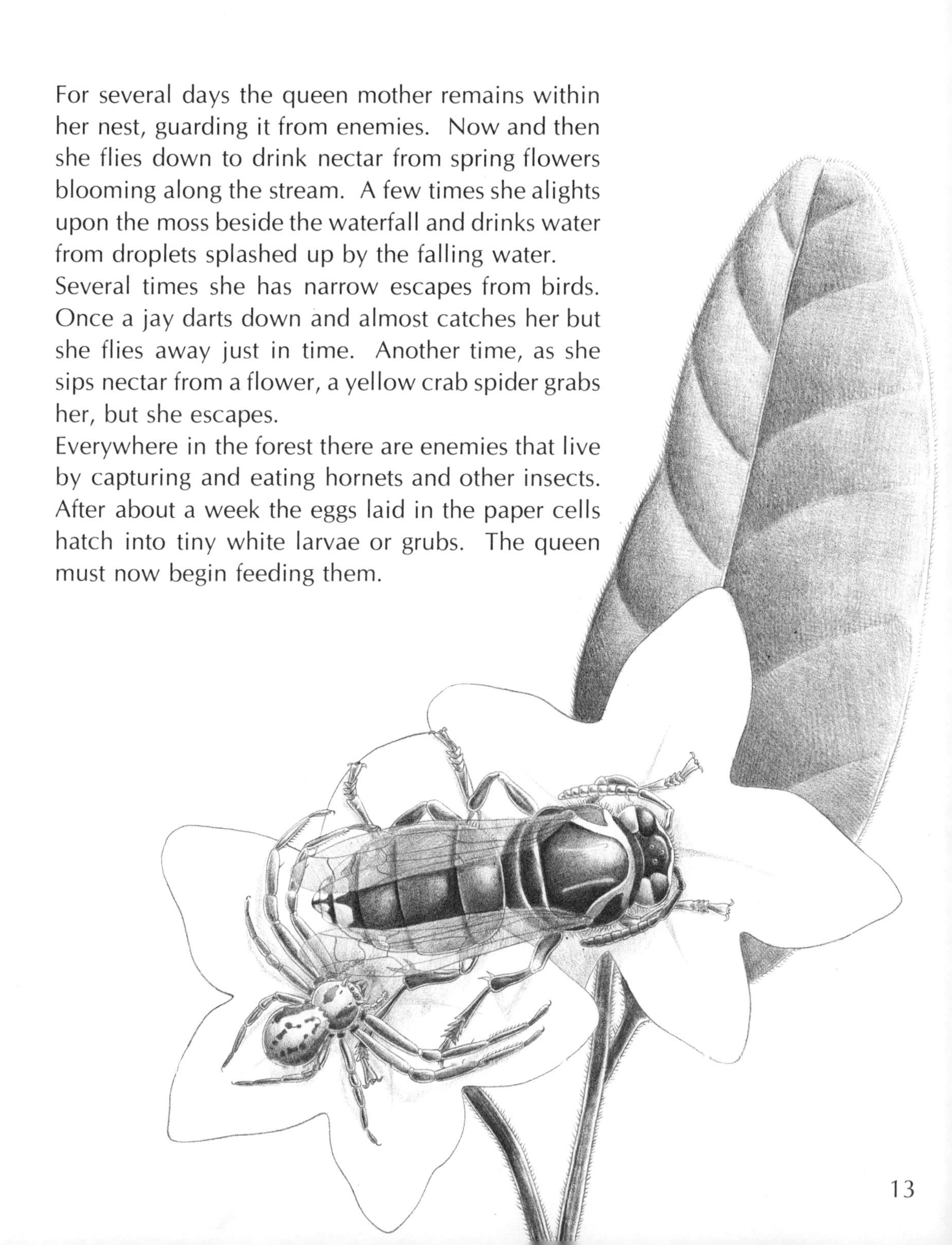

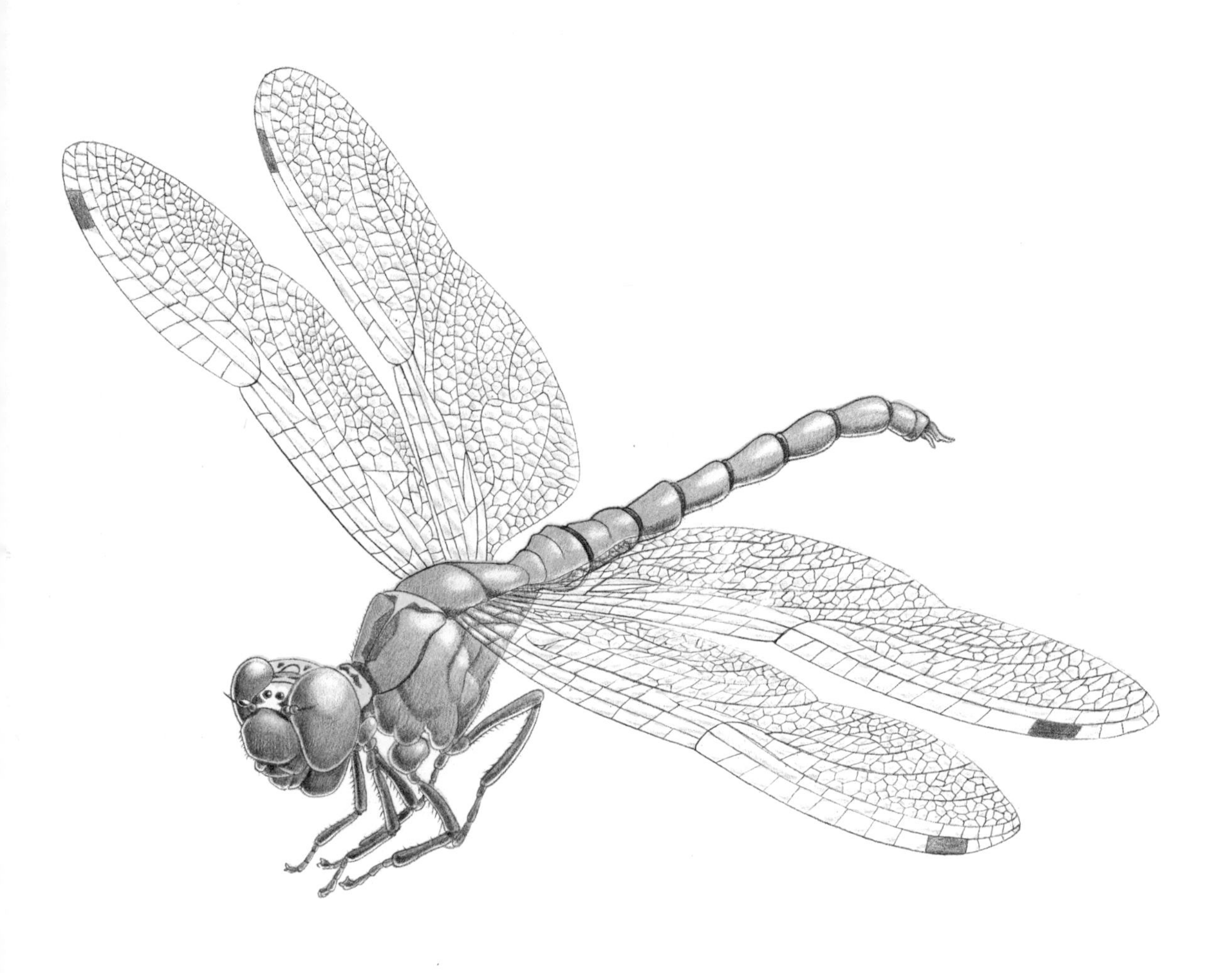

She flies up the stream seeking flies or other small insects. Her wings carry her above the waterfall beyond which is a large moss-covered boulder where she stops and rests for a few minutes. Nearby, a dragonfly sails across the stream. The dragonfly is also a hunter. It, too, is seeking flies but it is not interested in hornets.

A flycatcher perched high in a tree sees the queen and darts down. The queen escapes just as the bird's beak snaps shut with a loud click. The fly-catcher sails away and the queen flies on up the stream in search of flies.

It is mid-day and the stream is now bathed in the warm rays of the spring sun. Near the water's edge the queen spots a number of flies clustered on the damp sand. She flies down and captures one of them in her jaws. Then, perched on the twig of a blooming dogwood, she chews up the fly into a hamburger-like paste.

For a moment the queen remains on the twig, then sails back down the stream toward her nest. Arriving there she enters the nest's open bottom and begins feeding the hamburger-like paste to the small grubs in the cells. Then she flies back up the stream hunting for more flies.

Now that her young have hatched they need lots of food. So she must toil for long hours in the spring sun to keep them satisfied.

As the spring days pass the young hornet grubs continue to grow. As they grow larger, they need more and more food.

Each day the queen mother leaves the nest and hunts for flies. Sometimes she flies down the stream, at other times she hunts for game above the waterfall. Several times rains sweep across the mountains and force her to remain at home in the nest.

After about two weeks the larval hornets stop eating and spin silken cocoons around themselves. They also spin silken caps over the openings to their cells. After this they shed their skins and change into pupal hornets.

While the pupal hornets rest in their silken cocoons great changes take place in their bodies. Slowly their organs are being changed to fit them for flight. They now look like white ghosts of the adult hornets they will soon become.

Meanwhile, the queen guards her nest, leaving only occasionally to drink nectar from nearby flowers. Several times she is almost captured by jays, but each time she is lucky.

One afternoon a severe thunderstorm lashes the valley and hailstones as large as marbles beat many of the new leaves from the trees. One hailstone hits the side of the hornet nest and tears part of the paper wall away. The young hornets are not injured and, later, the sun comes out and warms the valley again.

One warm afternoon the pupal hornets, still in their cocoons, shed their skins for the last time and become winged adult workers. Shortly, they use their sharp jaws to cut away the silken caps to their cells and crawl out upon the comb to join their queen mother.

These first worker hornets are slightly smaller than the queen and smaller than workers that will be reared later. This is because the queen, working alone, has not been able to give them quite enough food.

For several days the young hornets cluster within the small paper nest with their mother who still feeds them what food she is able to gather. Instead of chewed-up flies she now brings them nectar collected from flowers.

The workers are nearly ready to take up their life's work. They will be collecting fiber for enlarging the nest and hunting food for the young that will soon be hatching from additional eggs laid by the queen.

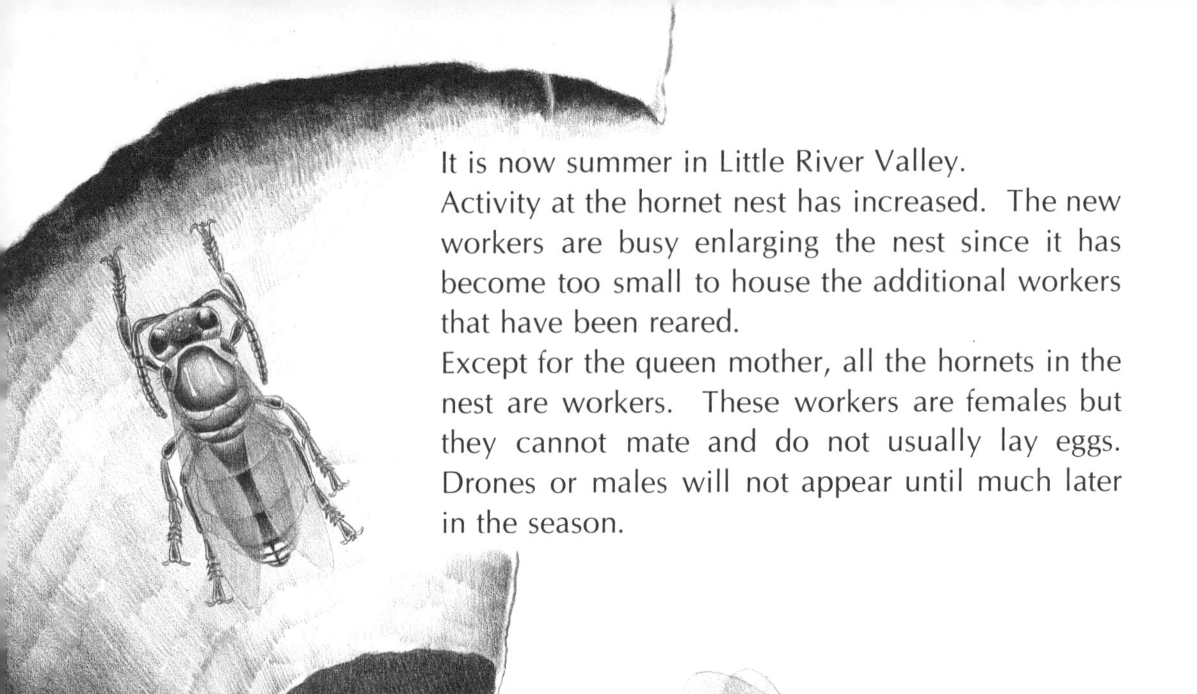

It is now summer in Little River Valley.

Activity at the hornet nest has increased. The new workers are busy enlarging the nest since it has become too small to house the additional workers that have been reared.

Except for the queen mother, all the hornets in the nest are workers. These workers are females but they cannot mate and do not usually lay eggs. Drones or males will not appear until much later in the season.

During the summer days the queen mother remains in the nest laying eggs in the new cells built by the workers. There is almost always a stream of hornets entering or leaving the nest. Some of them fly up the mountainsides hunting for flies. The captured insects are chewed into hamburger-like paste and carried back to the nest as food for the young in the cells. Other workers gather wood fiber for the making of paper. There are many tasks and the hornets are always busy.

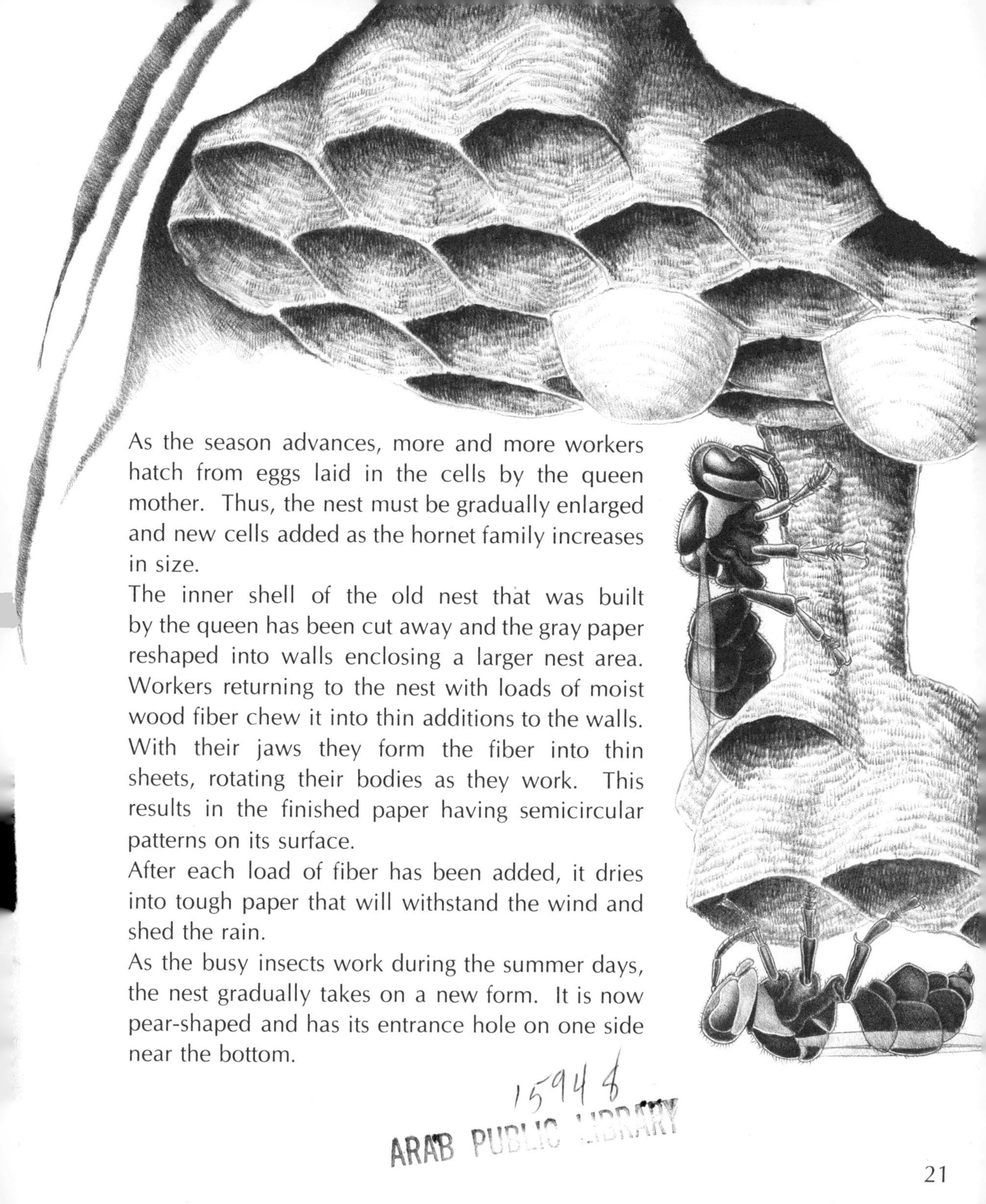

As the season advances, more and more workers hatch from eggs laid in the cells by the queen mother. Thus, the nest must be gradually enlarged and new cells added as the hornet family increases in size.

The inner shell of the old nest that was built by the queen has been cut away and the gray paper reshaped into walls enclosing a larger nest area. Workers returning to the nest with loads of moist wood fiber chew it into thin additions to the walls. With their jaws they form the fiber into thin sheets, rotating their bodies as they work. This results in the finished paper having semicircular patterns on its surface.

After each load of fiber has been added, it dries into tough paper that will withstand the wind and shed the rain.

As the busy insects work during the summer days, the nest gradually takes on a new form. It is now pear-shaped and has its entrance hole on one side near the bottom.

The hornet nest has also been completely changed inside. The small cluster of paper cells originally built by the queen is gone. In its place is a much larger comb containing nearly a hundred cells. Below this comb, other combs have been added, leaving only enough space between each story for the hornets to crawl about. The workers constantly feed and care for the young that are hatching from eggs the queen mother continues to lay in the additional cells.

Also, instead of a single paper wall enclosing the nest, the hornets have added several more layers of paper, leaving small air spaces between each wall. These air spaces serve to insulate the nest against the chill of the mountain nights and the heat of the summer days.

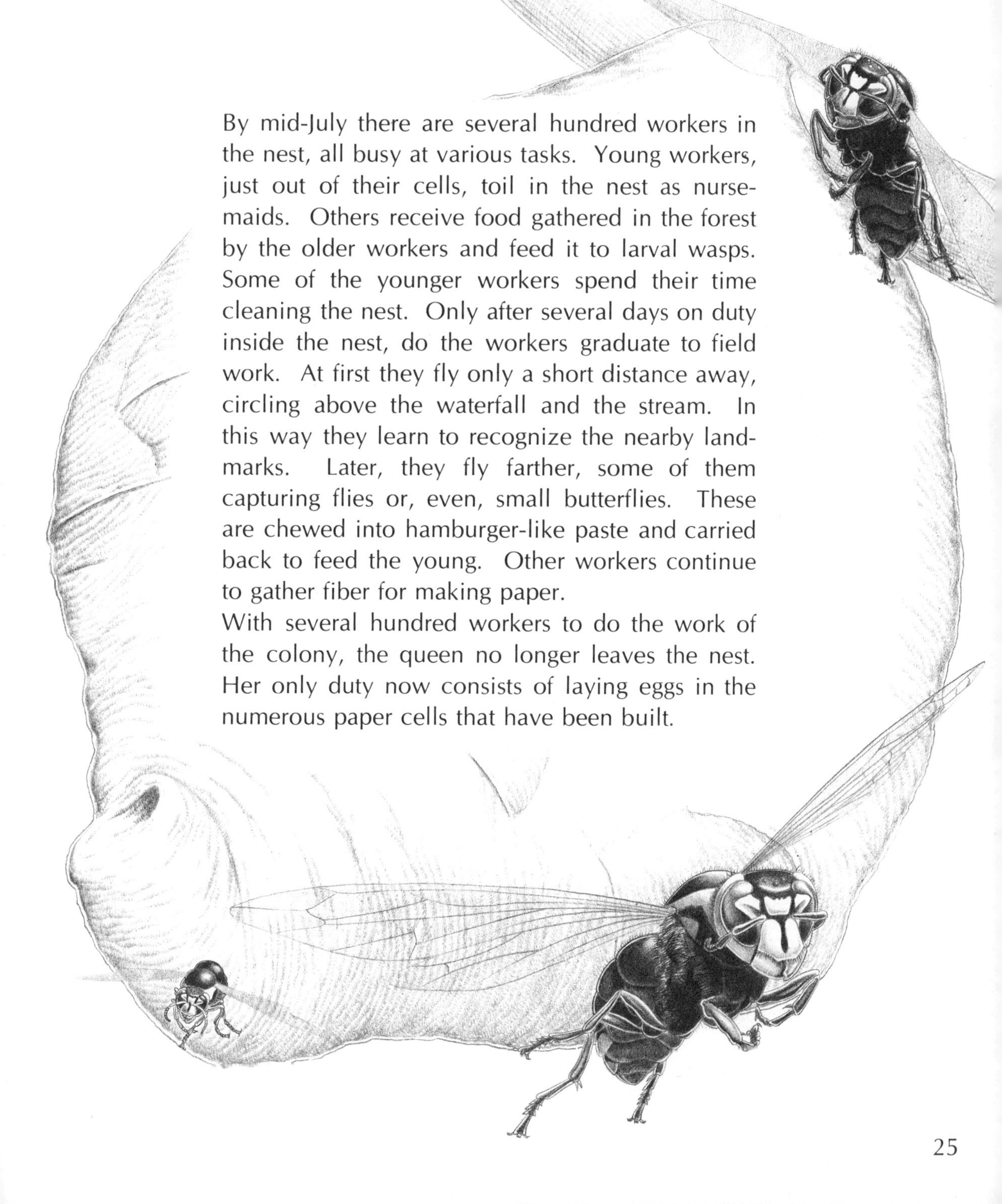

By mid-July there are several hundred workers in the nest, all busy at various tasks. Young workers, just out of their cells, toil in the nest as nursemaids. Others receive food gathered in the forest by the older workers and feed it to larval wasps. Some of the younger workers spend their time cleaning the nest. Only after several days on duty inside the nest, do the workers graduate to field work. At first they fly only a short distance away, circling above the waterfall and the stream. In this way they learn to recognize the nearby landmarks. Later, they fly farther, some of them capturing flies or, even, small butterflies. These are chewed into hamburger-like paste and carried back to feed the young. Other workers continue to gather fiber for making paper.

With several hundred workers to do the work of the colony, the queen no longer leaves the nest. Her only duty now consists of laying eggs in the numerous paper cells that have been built.

Summer days in the Great Smoky Mountains are often hot and the mountains are then blanketed each day by a blue haze. This smoke-like haze consists of fine droplets of oil breathed out of the dense vegetation and makes far-off mountains hard to see. This is the reason that the Indians, long ago, called them "The Great Smoky Mountains." The first white settlers who came to the area adopted the Indian name, but changed it into English.

Only after rains have cleared the air of the smoke-like haze can the more distant mountains be seen. The hornets who live in the nest above Huskey Falls cannot see the far-off mountains. Their eyes are not capable of seeing great distances.

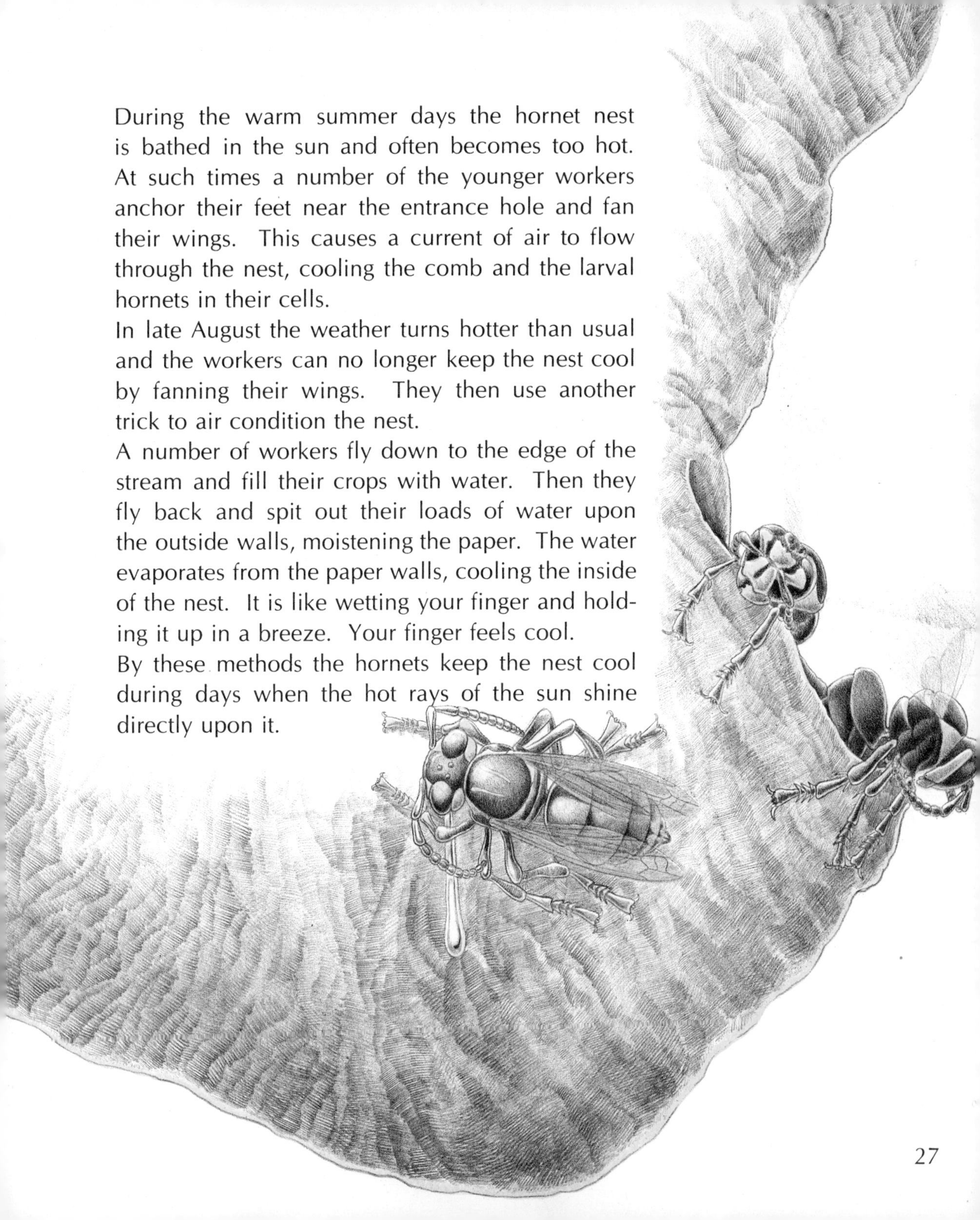

During the warm summer days the hornet nest is bathed in the sun and often becomes too hot. At such times a number of the younger workers anchor their feet near the entrance hole and fan their wings. This causes a current of air to flow through the nest, cooling the comb and the larval hornets in their cells.

In late August the weather turns hotter than usual and the workers can no longer keep the nest cool by fanning their wings. They then use another trick to air condition the nest.

A number of workers fly down to the edge of the stream and fill their crops with water. Then they fly back and spit out their loads of water upon the outside walls, moistening the paper. The water evaporates from the paper walls, cooling the inside of the nest. It is like wetting your finger and holding it up in a breeze. Your finger feels cool.

By these methods the hornets keep the nest cool during days when the hot rays of the sun shine directly upon it.

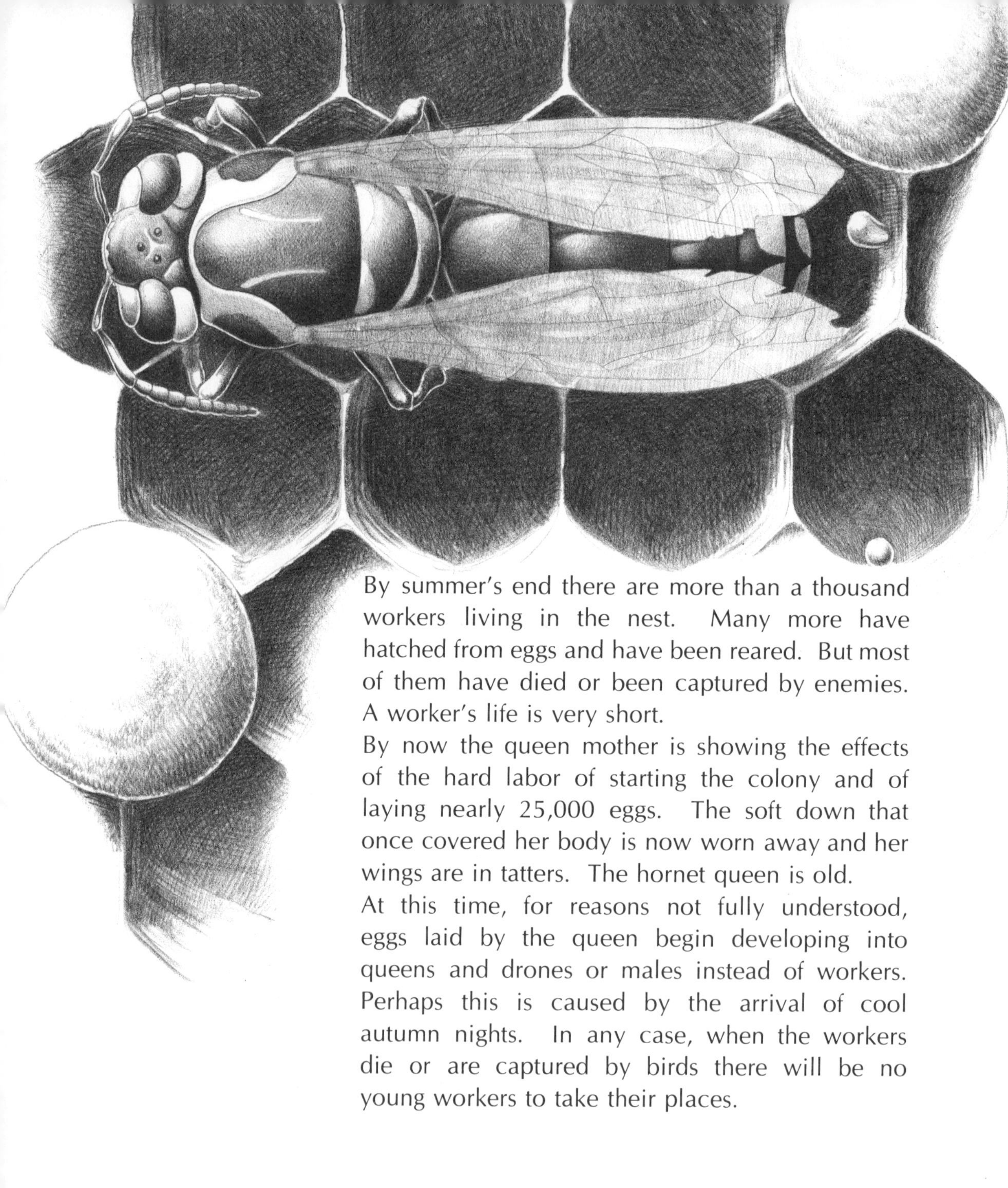

By summer's end there are more than a thousand workers living in the nest. Many more have hatched from eggs and have been reared. But most of them have died or been captured by enemies. A worker's life is very short.

By now the queen mother is showing the effects of the hard labor of starting the colony and of laying nearly 25,000 eggs. The soft down that once covered her body is now worn away and her wings are in tatters. The hornet queen is old.

At this time, for reasons not fully understood, eggs laid by the queen begin developing into queens and drones or males instead of workers. Perhaps this is caused by the arrival of cool autumn nights. In any case, when the workers die or are captured by birds there will be no young workers to take their places.

It is now mid-September in the Great Smoky Mountains. Only a few of the forest trees have, as yet, been touched by Jack Frost's magic paint brush. The leaves of some sourwoods and sumacs have just begun to turn scarlet.

Winter and its cold will soon arrive in the valley. In the meantime the hornet workers toil on as usual. But they now hunt only for food to feed the larval males and queens in the cells. The workers no longer collect wood fiber to enlarge the nest.

Today they stream out of the entrance, working as if the warm weather will last forever. They have no way of knowing that winter will soon arrive and that they will all be dead.

As the autumn days pass, the nights grow colder. Everywhere the mountains are now splashed with color and each day the leaves of the oaks and maples become more brilliant. In time the leaves begin breaking away from the twigs and spiraling down to the ground, gradually covering it with a multicolored carpet.

Near the waterfall there is a large boulder beneath which a groundhog has dug its den. Each day it comes out and suns itself on the top of the boulder. The groundhog has become fat and will soon stop feeding. Then it will curl up in its den where it will sleep the winter away. Chipmunks are busy in the surrounding forest gathering nuts and carrying them into their underground homes.

Every creature of the forest, in its own way, is preparing for the coming winter.

One by one the drone and queen larvae in their paper cells become full grown. They then spin silken cocoons around themselves and change into pupae.

No longer finding it necessary to gather food for the larvae, the worker hornets buzz about the nest. Some of them wander away through the forest, dropping down now and then to sip nectar from goldenrods or asters.

The once-active life in the hornet nest suspended from the grapevine above Huskey Falls is gradually coming to a halt. Time is running out.

By late October most of the trees on the mountains are nearly bare of leaves. Each breeze causes more of them to fall to the earth. Many leaves have fallen into Huskey Branch and now and then a gayly colored leaf tumbles down over the waterfall and into the pool. It floats about in the current for a time, then drifts on down the river. On some mornings white frost can be seen on the tops of the higher mountains.

Today the sun has warmed the hornet nest above Huskey Falls and the young queens and drone adults rest on the combs inside it. Among them are a few workers but they seem uncertain what to do. There is now no work for them.

Inherited knowledge or instinct now causes the queens and drones to leave the nest. One by one they fly out of the entrance and buzz up the mountainside or across the river.

In the meantime, queens and drones from other hornet nests in the valley have also felt the urge to leave their paper homes and have flown away through the forest.

In time the hornet queens meet drones and mating occurs. These are the hornets' honeymoon flights and when they are over the drones wander away and die or are captured by birds.

After mating, the young queens fly about for a time, but some instinct tells them that they must seek places to spend the winter. They explore rotten logs and stumps. A few of them crawl under stones or fallen bark. It is most important that the place they find be safe from the cold.

Many of the mated queens from the nest above the falls fly up the stream investigating likely places. One crawls into an old beetle tunnel in a rotten log; another finds a crack in the bark of a still-standing oak. One queen, more ambitious than her sisters, flies all the way up to a point where an old logging trail crosses the stream. Here she finds a dead oak tree, its great trunk resting upon the ground. Part of its bark has fallen away, exposing the bare wood beneath it.

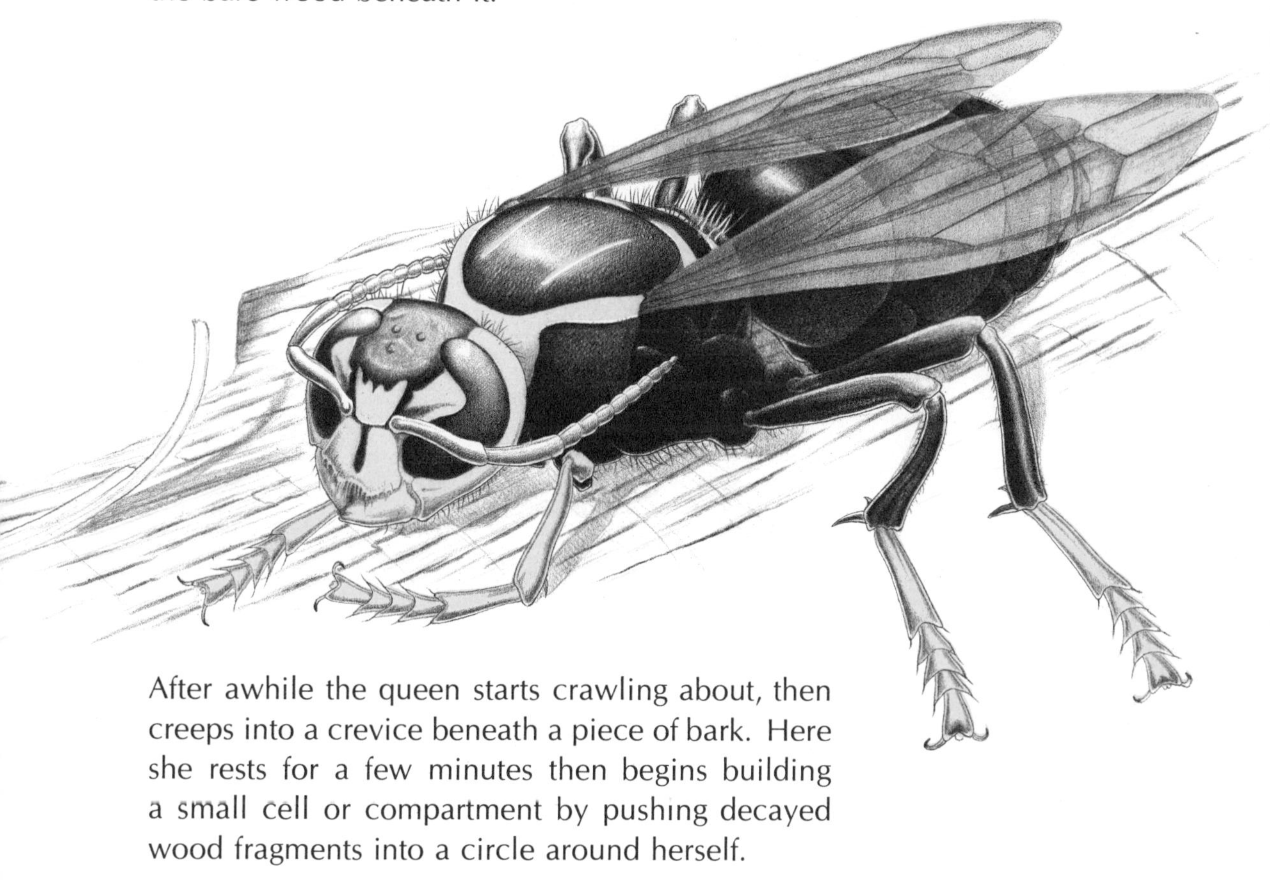

After awhile the queen starts crawling about, then creeps into a crevice beneath a piece of bark. Here she rests for a few minutes then begins building a small cell or compartment by pushing decayed wood fragments into a circle around herself.

The queen buzzes down and alights upon the log. The autumn sun is warm and the queen rests quietly upon the exposed wood for awhile. Nearby, a chipmunk scurries about in search of acorns but it pays no attention to the hornet queen. It is not interested in insect food.

A few feet away, the stream ripples down through the forest, dropping over boulders and spilling into small pools. Beside one pool a skunk has found an underground yellow jacket nest and is now busily digging into it and eating the insects.

All this, of course, is part of the normal woodland scene. So the queen is not alarmed and continues to rest quietly upon the oak log.

Having made a cleared space surrounded by walls of wood fragments, the queen settles down in the center and tucks her antennae and wings under her legs. Then she rests quietly. She has now done all she can to prepare herself for the coming winter.

All the queen's actions have been through knowledge or skill inherited from past generations of hornets. She does not know that winter is coming, but she responds to age-old instincts that will help her to survive the cold months.

By November most of the hornet queens in the valley of Little River have found places to hibernate. Some were unlucky; they were captured by birds. That, however, is Nature's way of balancing hornet populations. Many more queens were produced in autumn than will be needed. If all of them were to survive until spring and build nests there would be far too many hornet nests and not enough food to feed all the young that would be reared in them.

Always both animals and plants produce more young or seeds than can possibly survive. In this way most living creatures are able to live to reproduce their kind.

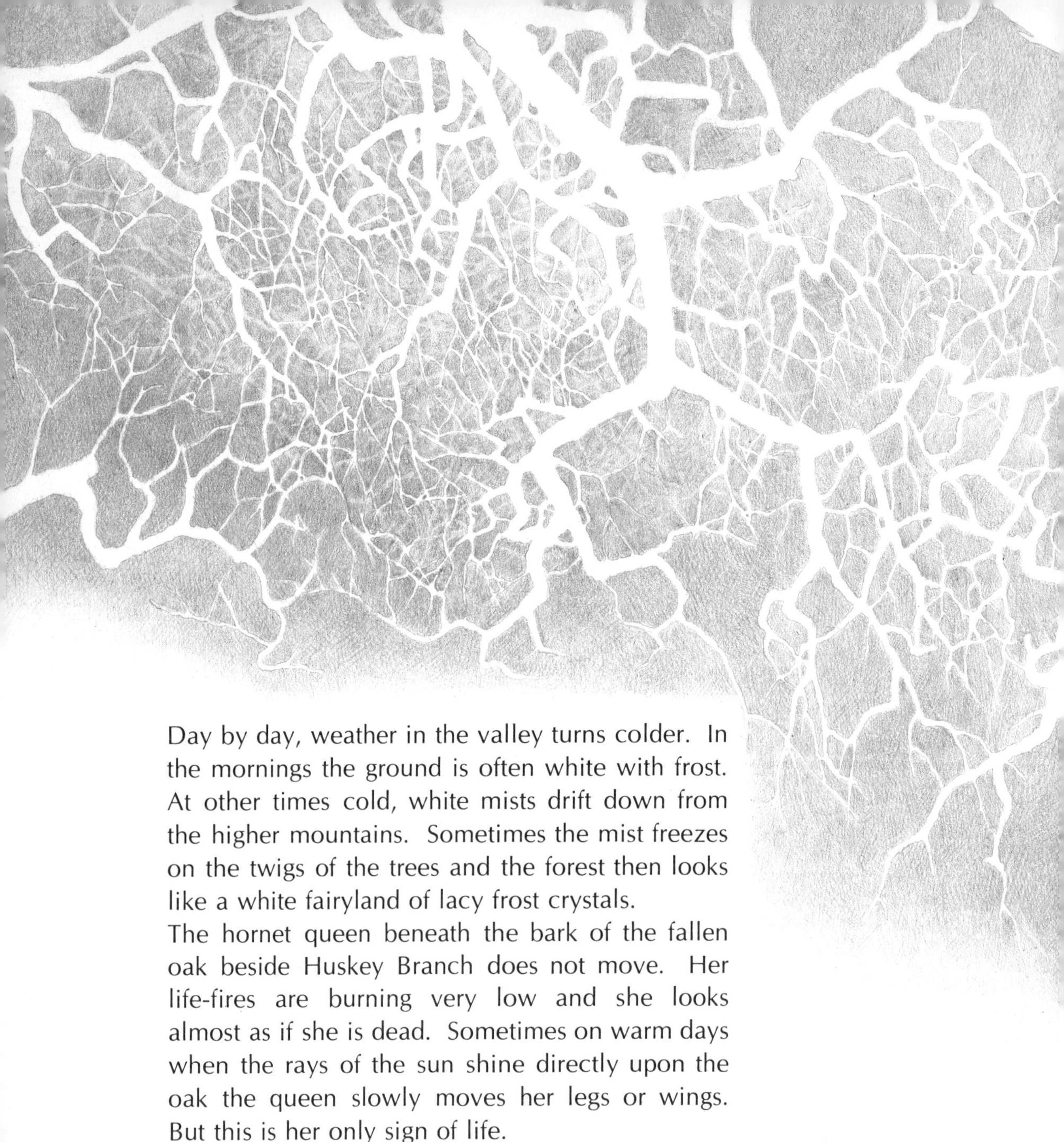

Day by day, weather in the valley turns colder. In the mornings the ground is often white with frost. At other times cold, white mists drift down from the higher mountains. Sometimes the mist freezes on the twigs of the trees and the forest then looks like a white fairyland of lacy frost crystals.

The hornet queen beneath the bark of the fallen oak beside Huskey Branch does not move. Her life-fires are burning very low and she looks almost as if she is dead. Sometimes on warm days when the rays of the sun shine directly upon the oak the queen slowly moves her legs or wings. But this is her only sign of life.

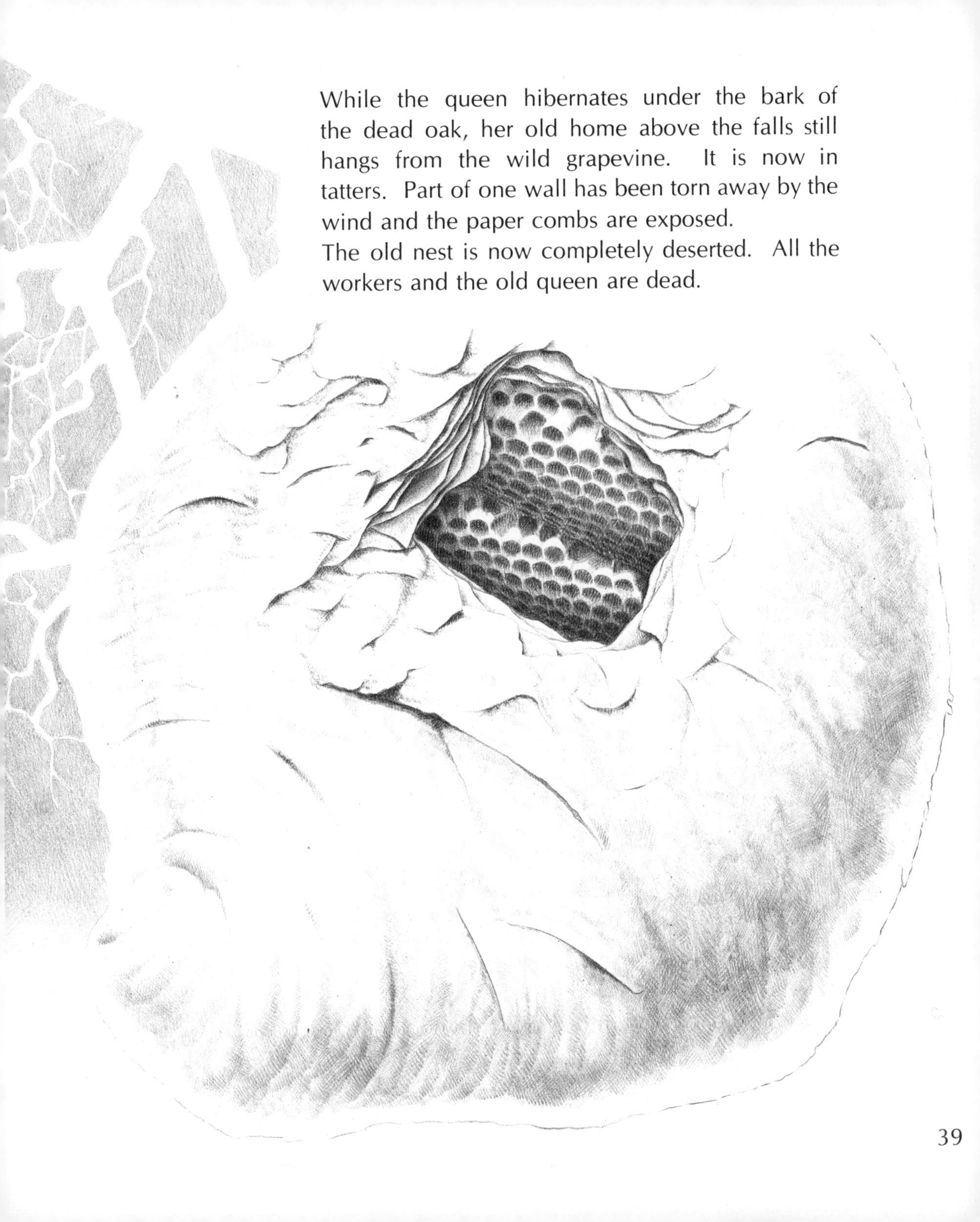

While the queen hibernates under the bark of the dead oak, her old home above the falls still hangs from the wild grapevine. It is now in tatters. Part of one wall has been torn away by the wind and the paper combs are exposed.
The old nest is now completely deserted. All the workers and the old queen are dead.

One afternoon a woodpecker flies down and perches on the grapevine near the nest. It hops along the vine and peers at the exposed combs, then flutters its wings and comes to rest upon the nest. Then it begins hammering upon the comb with its sharp beak. The woodpecker soon discovers that there is no food in the nest. So it spreads its wings and flies away up in the mountainside seeking dead trees where fat beetle grubs can probably be chiseled out of the wood.

As the days pass, winds continue to damage the old hornet nest. More fragments of its paper walls are torn off and flutter away.

One day in December a crow spots the nest and flies down and lands upon it. But the weight of the bird causes the nest to break away from the vine and fall.
The crow is alarmed and flies away down the river.
The hornet nest falls into a pool below the waterfall where it drifts about in the current, then is carried on down the stream and into the river. A trout sees the paper nest and swims over to look at it, then glides away. Soon the old hornet nest floats on down the river and disappears.
This is the end of the paper nest so carefully built during the summer days by the industrious hornets. It sheltered them from both heat and cold and protected them from enemies.
The nest is gone but it served its purpose well.

The month of January begins with a great storm that sweeps across the mountains and down into the valley. During the night the temperature drops to near zero and then it begins to snow.

By dawn the valley is buried under more than a foot of fresh snow and the sound of the river can no longer be heard. The water now flows beneath a thick covering of ice.

The snow is like a soft blanket over the fallen oak tree beside Huskey Branch but the hibernating queen is safe. The snow and the bark of the dead tree protect her from the cold.

While the forest may seem deserted, beneath the deep snow creatures of several kinds are active. Most active of all the small creatures under the snow blanket are the shrews. They are smaller than mice and have needle-sharp teeth, well fitted for capturing and eating insects. Sometimes they even prey upon mice even though the mice are larger than they are.

The shrews discover many hibernating insects and devour them. A shrew needs lots of food.

The hornet queen beneath the dead oak beside Huskey Branch is lucky. The shrews do not find her and so she sleeps on, knowing nothing of the cold or snow or enemies.

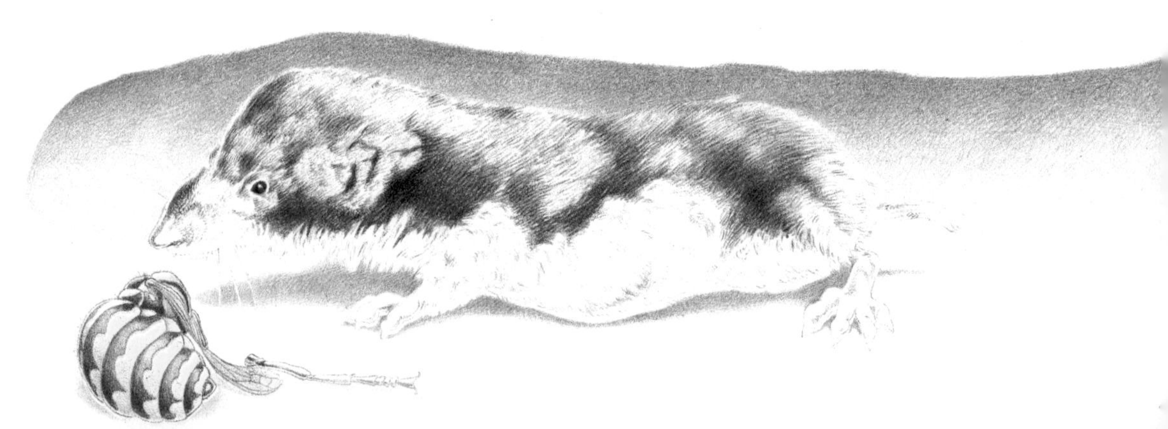

And so the winter months pass away. In spring the snow melts and the water rushes down the valley in great torrents. Boulders tumble about in the river's bed carrying large tree trunks along in its swift current.

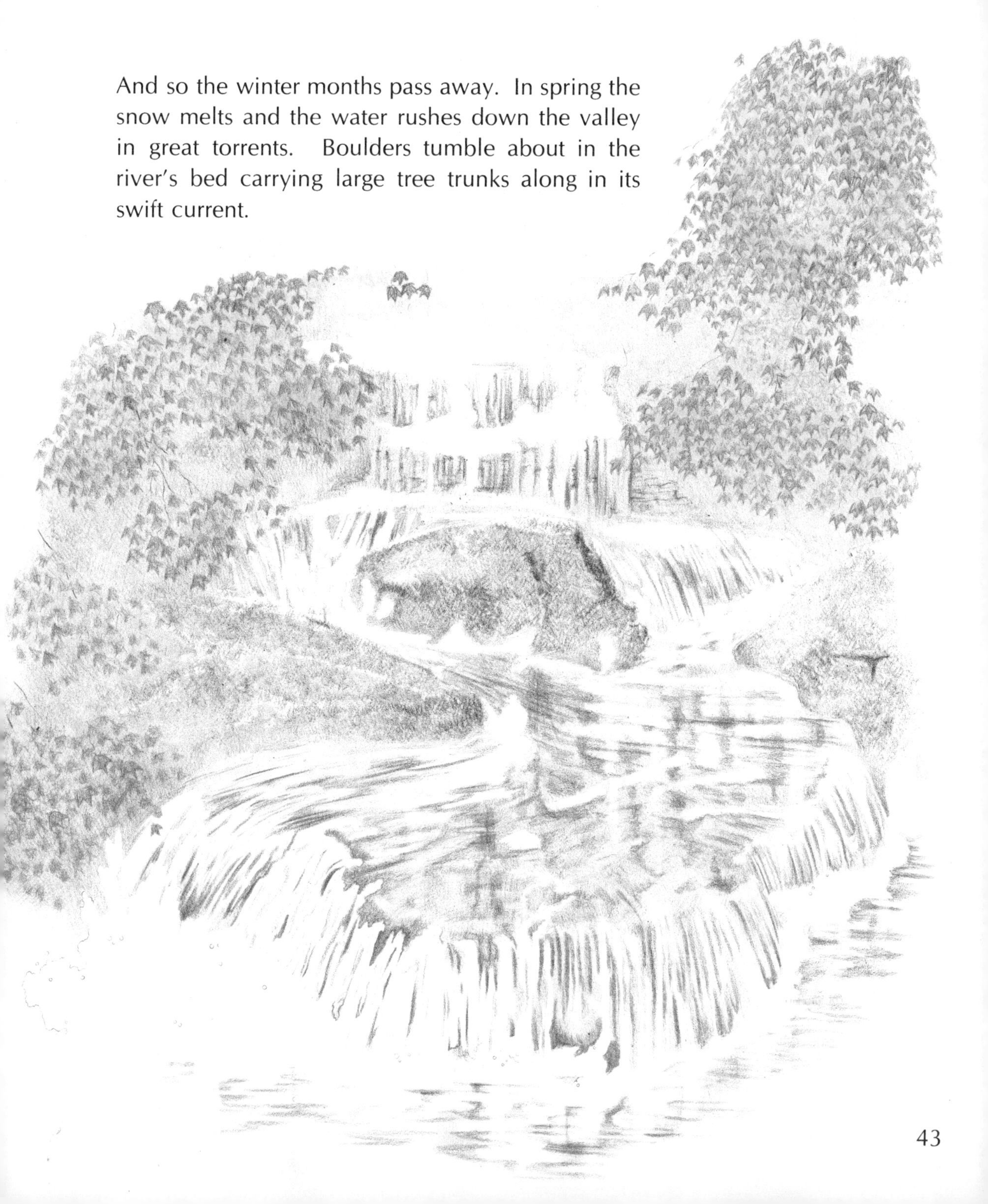

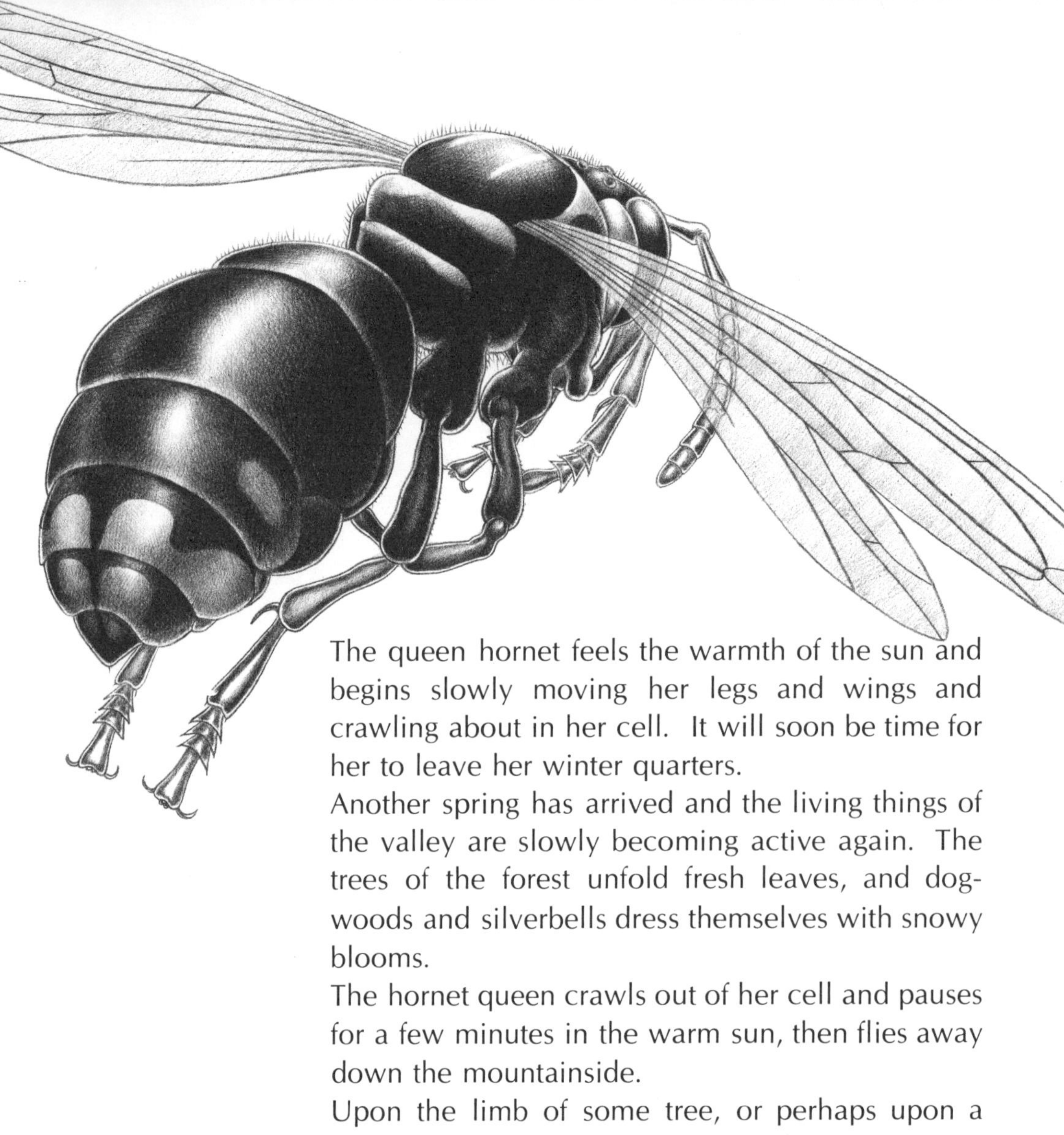

The queen hornet feels the warmth of the sun and begins slowly moving her legs and wings and crawling about in her cell. It will soon be time for her to leave her winter quarters.

Another spring has arrived and the living things of the valley are slowly becoming active again. The trees of the forest unfold fresh leaves, and dogwoods and silverbells dress themselves with snowy blooms.

The hornet queen crawls out of her cell and pauses for a few minutes in the warm sun, then flies away down the mountainside.

Upon the limb of some tree, or perhaps upon a wild grapevine, the queen will build a paper nest. She will start another hornet colony where workers will toil through the days of summer just as a million past generations of hornets have done before.

SCIENTIFIC NOTES

Insects, in many cases, have been employing techniques that antedate by millions of years similar ones developed by man. For example, dragonfly nymphs were darting through the water by jet-propulsion during the Carboniferous Period at least 300-million years ago. Insects were also the first creatures to fly; they were speeding through the air millions of years before birds arrived on the world scene. Of all insects' achievements, that of powered flight was without doubt their greatest evolutionary step. In addition, certain insects, such as bees and ants, long ago developed the ability to navigate or find their way by the sun. The manner by which these insects employ the principles of solar navigation to find their way is one of the true marvels of science, and they were using this technique long before man learned how to find his way across the trackless seas by observing the heavenly bodies and their movements.

Insects also entered the field of agriculture long before man began cultivating food crops. Several kinds of ants, termites and beetles raise fungi as sources of food and may, thus, be considered to be keepers of mushroom gardens.

It is in the building trades, however, that many insects excel and their skills were evolved a very long time ago. Caddis insects of several kinds build cases of stones cemented together and so were the original stone masons. Mason wasps — there are many kinds — fashion cells of clay in which their young are reared. Termites, also, build clay nests. A number of insects, including certain tropical ants as well as our native leaf-rolling crickets, construct nests by cementing leaves together and so were the first tent-makers. Insects were also the first creatures to make paper. Certain ants build paper-carton shelters in which their aphid "livestock" is housed and protected. The most remarkable of all paper-making insects, however, are the hornets and yellow jackets that fabricate paper nests out of wood fiber collected from dead trees, fence posts or weathered boards. The paper produced by these insects is of excellent quality, being formed in thin sheets that resist weathering by wind and rain for considerable periods. Pieces of this paper may even be written or typed upon.

Many of the most skillful paper-making hornets live in tropical lands but here in our own country are found several that are very adept at the art of making paper. One of these is the common bald-faced hornet (*Dolichovespula* (*Vespula*) *maculata*). This is a large black hornet with yellowish-white markings on its body and having its face marked with patterns of a similar color. Its jaws are very sharp and powerful, well fitted for capturing prey and for forming plant fiber into paper-thin sheets. The nests built by these industrious insects are often a foot or more in diameter and, when finished, may have walls made up of fifteen or more layers of paper. It ranges throughout the United States.

The bald-faced hornet is the insect of our story which is laid in the Great Smoky Mountains of Tennessee where the writer has often observed them at work and studied their amazing social organization. Near summer's end it is always with a feeling of sadness that the toiling hornets are watched at their work. Heedless that with the coming of late autumn all the workers will be dead and the once-populous nest deserted, they labor as though summer will last forever. During the long, hot days they stream out of the nest suspended from a grapevine above the falls of Huskey Branch and fly away through the dark forest. Other workers are flying back to the nest carrying food or loads of plant fiber. By late October the nest is deserted; not a single hornet remains. Queens, produced in early autumn, have by this time found hibernating places beneath logs or under the bark of dead trees and will remain there all winter.

In a typical hornet colony, depending on the season, there may be three kinds of individuals or castes; workers, queens, and males or drones. During early summer there is only one queen mother and no drones. Workers and queens are both females and arise from fertilized eggs. The difference between workers and queens results, probably, from food they receive during their larval stages. Drones or males, on the other hand, arise from unfertilized eggs and thus have no fathers. During early summer there is but one egg-laying queen but as the season advances a number of young queens and drones emerge and the number of workers produced begins to decrease. At about this time the wonderfully organized social structure of the colony starts breaking down and the new queens and drones soon leave the nest and mate. After mating the drones die but the mated queens find suitable places to hibernate.

After all the hornets are gone the old paper nest is gradually torn apart by winter storms or by the activities of inquisitive birds. Sometimes it falls to the ground. Thus, of all the hornets that once lived in the nest during summer, only the mated queens remain alive. In spring they will emerge from their hibernating places and start new hornet colonies. They are the hornets' only link between the seasons.

INDEX

ABOUT THE AUTHOR AND ARTIST

Ross Hutchins' poetic style brings out the special qualities in the daily world of an insect. His interest and travels in the Great Smoky Mountains have supplied him with authentic background material for several of his recent books including PAPER HORNETS.
He is the author of more than 30 books on nature studies as well as numerous articles in Natural History, National Geographic, and other journals. Dr. Hutchins holds a Ph.D. from Iowa University and is listed in Who's Who and American Men and Women of Science. He is also a member of the Appalachian Club.

Peter F. Zallinger's precise and skillful style adds to the authenticity of PAPER HORNETS. Although this is one of his first illustrated children's books, Mr. Zallinger is well acquainted with this art form, having been raised in a family of accomplished illustrators.
He is a graduate of Yale University, and he and his wife presently live in Brookline, Massachusetts.